P9-DEB-704

FIGHTING FORCES ON LAND

# M113 ARMORED PERSONNEL VEHICLE

DAVID BAKER

Rourke
Publishing LLC
Vero Beach, Florida 32964

www.rourkepublishing.com

PHOTO CREDITS: All photos courtesy United States Department of Defense, United States Department of the Army

Title page: *The advantage of tracks over wheels; a M113 APC hauls a Humvee to dry ground, its wheels stuck fast in the mud.*

Editor: Robert Stengard-Olliges

Library of Congress Cataloging-in-Publication Data

Baker, David, 1944-
Armored personnel vehicle / David Baker.
p. cm. -- (Fighting forces on land)
Includes index.
"Further readings/websites"--p. 32.
ISBN 1-60044-247-1
1. M113 (Armored personnel carrier)--Juvenile literature. 2. United States--Armed Forces--Armored troops--Juvenile literature. I. Title. II. Series.
UG446.5B234 2007
623.7'475--dc22

2006010784

Printed in the USA

CG/CG

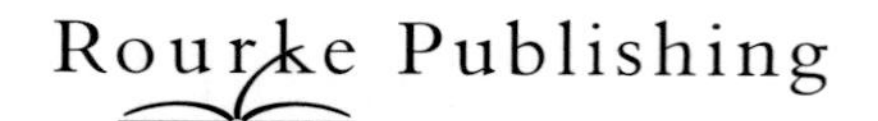

www.rourkepublishing.com – sales@rourkepublishing.com
Post Office Box 3328, Vero Beach, FL 32964

# Table of Contents

# THE NEED FOR SPEED

CHAPTER ONE

Battles are won or lost for many different reasons but one of the most important to avoid is the lack of transport to get the troops where they are needed – fast! In the last century, the development of motorized transport was a vital part of that effort and military vehicles were designed for many specific roles.

▲

*The rear entrance door affords some limited armored protection for a squad of infantry and allows rapid exit and entry.*

▲

*Protection for the infantry is important. Lives depend on adequate cover. This infantry squad really needs more armor protection.*

▲

*Hunkered down behind their M113 armored personnel carrier, these soldiers are protected by the light armored skin of the vehicle and by their ability to withdraw at will or move up on the enemy positions.*

As warfare moved from set-piece battles, involving armies lining up opposite each other, to large-scale conflict carrying troops through urban as well as **rural** environments, the need for **specialized** vehicles increased. Some, however, were needed that could adapt quickly to a variety of different roles and requirements and such a vehicle is the M113 **armored** personnel carrier, or APC.

CHAPTER TWO

# MOBILITY

*Traveling at speed across a dirt road, the M113 Gavin optimizes the modern, mobile, infantryman backed up with firepower and force. Note the Humvee in the background.*

Troops are equipped with tanks, light armored fighting vehicles, and tracked vehicles for crossing rough country. Mobility – the ability to carry people quickly across difficult or undulating **terrain** – is an important part of keeping up with the fighting vehicles. Infantry must be picked up and carried around the battlefield or be taken to important spots where they can be used on foot or to hold defensive positions.

▲
*Infantrymen patrol against insurgents around streets.*

**Flexibility** is often the difference between losing and winning. In World War II (1939-1945) infantry lacked proper vehicles to move around in and would frequently hitch a ride on the exterior surface of tanks as a means of moving quickly from one place to another, many being killed when the tanks came under attack.

▲

*Gavin APCs transport troops safely into and out of downtown areas for sunset patrols.*

Lessons learned there were quickly applied to the needs of US forces facing new and imposing threats in an age of increasing **tension**. Troop vehicles were needed that could not only prove their mobility on the battlefield but in their flexibility for different roles as well.

▲

*The adaptable Gavin is ideally suited for street patrol where foot soldiers frequently disperse crowds or carry out police-keeping duties.*

# Designing for Protection

CHAPTER THREE

After World War II US forces faced possible attack from the Soviet Union or communist China. In the Korean Conflict (1950-1953) the US and its allies discovered the need for a new range of modern weapons and equipment resulting in the M59 and M75 armored personnel carriers designed by Ford. Improvements led to the M113 designed in the early 1960s as the US Army's first modern "battle taxi" capable of transporting infantry forces around the mechanized battlefield providing protection as well as mobility.

▲

*Machine gun armor shield kits are available for the commander's cupola. Rear shields each allow for a pintle-mounted gun.*

▲

*The Gavin is readily transported by rail or air. Here a convoy of APCs roll off a freight car.*

The design of the APC is a **compromise** between the need for it to be as strong and rugged as possible yet light and simple enough to be carried by air and easily maintained using few tools and little support equipment. The M113 strikes a balance by incorporating aircraft grade aluminum materials in its construction with approximately the same strength as steel and, because that makes it lighter, a small 209 hp two-stroke **diesel** engine for simplicity.

Popularly known as the Gavin, the M113 carries a driver and track commander plus 11 infantrymen at a top speed of 37 mph with a range of about 200 miles. It came with a standard 0.5 **caliber** machine gun for defense or for attacking lightly defended enemy positions.

▲

*Armed with a 0.5 caliber M2 and a single 7.62 mm machine gun, the exterior of the Gavin makes an ideal place for hanging kit bags. Note the external fuel can and the unique shape of the door aperture.*

▲

*APCs operate as a functional part of integrated forces using helicopter support and ground troops.*

# ADAPTING TO NEW ROLES

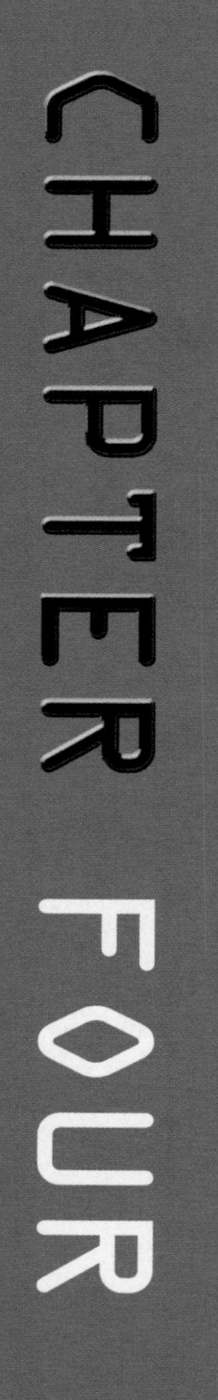

▲

*Adapted as a TOW anti-tank missile launcher, the Gavin is designated M901 and has lightly armored sides.*

Soon after initial production of the Gavin began in 1960 the Army started finding new roles for the APC that could take the basic chassis and track system and **modify** them for different tasks. During the Vietnam War of the 1960s, the Gavin got a thorough workout and adaptations for local needs were numerous.

▲

*With interior modifications and a specially designed hard shell, the Gavin can double as a command vehicle.*

Capable of putting troops into environments difficult for other vehicles to reach the M113 gave infantry an opportunity to move in and out fast and with some protection. The Gavin evolved into an armored medical evacuation vehicle, missile launcher, cargo carrier, self-propelled mortar, command post, and smoke generator.

▲

*Medical evacuation vehicles are deployed throughout the theater and that means they must be carried quickly across great distances, especially in war zones like Iraq.*

▲

*Ideally suited for speedy access to all areas of the battlefield, the Gavin has performed well as a MEDEVAC vehicle collecting wounded for the field hospital.*

▲

*Astronauts ride the NASA standby rescue vehicle built up on the M113 chassis and deployed to the Kennedy Space Center.*

*Only relatively minor changes are necessary to adapt the Gavin to a launch vehicle for 81 mm or 107 mm mortars in which configuration the vehicle is designated M125 and M106 respectively.* ▶

Some were modified as armored flamethrowers, light or medium **reconnaissance** vehicles, and a few were fitted with armored turrets and a 76 mm gun from the Saladin armored car to carry out a fire support role. In all, more than 40 different variants and numerous unrecorded field modifications make the Gavin one of the more adaptable vehicles ever used by armies in the field. The Gavin has not only been adopted by the military, several M113 chassis being used as the basic building block for duties where fire and blast protection is essential. Although it has never been used, NASA has a modified M113 as a standby rescue vehicle for astronauts at the Kennedy Space Center fleeing the Shuttle in emergencies.

# IMPROVING THE DESIGN

CHAPTER FIVE

The first major improvement to the basic design was the M113A1 that appeared in 1964 followed by the A2 in 1979 and the A3 in 1987. The latest A3 version has a more powerful 275 hp six-cylinder diesel engine, can reach 41 mph, has greater range with strap-on fuel tanks, and can carry additional armor for added protection.

▲

*Engine air filters and environmental protection is never more necessary than for APCs operating in hot desert regions. Dust storms can develop in minutes and create enormous surges to clog machinery and reduce visibility.*

It can also carry additional firepower for defense. Applying lessons from operations around the world in a wide variety of climates and local conditions the Army has made changes to keep the Gavin up to date with modern developments. Not only in technology but also on the battlefield.

Only the M113A3 has the speed and the maneuverability to keep up with the M1A1 Abrams main battle tank and the M2/M3 Bradley fighting vehicles so that has become the standard specification, replacing the older and slower versions.

Partly concealed by an armor protection kit the M2 machine gun affords protection to a convoy of Gavins in Iraq.

# CHAPTER SIX

# LEGACY

More than 80,000 M113s have been built since the early 1960s, many having been exported to more than 50 countries and the Gavin can be seen around the world in almost every hot spot where infantry is needed.

▲ *The US infantryman whose safety and in whose support the Gavin was built and operated.*

*Global deployments are made for disaster relief as well as war when international efforts bring together forces from different countries. Here a Russian AN-124 prepares to airlift US vehicles to a relief zone.*

Improvements are being made all the time and modifications to existing as well as upgraded vehicles will include digital communications and control systems, satellite navigation, and better means of protecting the crew. Power is also available for laptop warfare where infantrymen control external remote devices from the protection of their M113.

The M2 and M3 Bradley fighting vehicle are gradually replacing the M113 but the aging Gavin still has a long life ahead and new **roles** to perform. Today, US forces are called upon to operate in a wide variety of environments and in different types of conflict. The adaptable M113 has changed over the last several decades to support our troops wherever they may go.

# Glossary

**armor** (AR mur) – strong metal protection for military vehicles

**caliber** (CAL uh ber) – the diameter of the barrel of a gun

**compromise** (KOM pruh mize) – to agree to accept something that is not exactly what you wanted

**diesel** (DEE zuhl) – a fuel used in diesel engines that is heavier than gasoline

**flexibility** (FLEK suh buhl uh tee) – being able to adapt to new, different or changing conditions

**modify** (MOD uh fye) – to change something slightly

**reconnaissance** (re KUHN uh senss) – the active gathering of information about an enemy, or other conditions, by physical observation

**role** (ROHL) – the job or purpose of a person or thing

**rural** (RUR uhl) – to do with the countryside or farming

**specialized** (SPESH uhl lized) – focused efforts on in a special activity, field,or practice

**tension** (TEN shuhn) – the tightness or stiffness of a rope, wire, or thing

**terrain** (tuh RAYNE) – the physical features of the land or ground

## INDEX



## FURTHER READING

Green, Michael and Gladys. *Armored Personnel Carriers: The M113*. Capstone Press, 2005
Verlinden, Francois. *Warmachines No2: M113*. Verlinden Productions, 1990

## WEBSITES TO VISIT

http://www.fas.org/man/dod-101/sys/land/m113.htm
http://www.army.mil/fact_files_site/M113.htm

## ABOUT THE AUTHOR

David Baker is a specialist in defense and space programs, author of more than 60 books and consultant to many government and industry organizations. David is also a lecturer and policy analyst and regularly visits many countries around the world in the pursuit of his work.